The Bunny Trail to Enlightenment

By Kimberly Kingsbury

COPYRIGHT PAGE

Publisher's Information goes here.

THE BUNNY TRAIL TO ENLIGHTENMENT

Copyright © 2018 Kimberly Kingsbury

Designer: Ian Kirkpatrick, www.irishbrush.artstation.com

Editor: Jessica Bryan, www.oregoneditor.com

ISBN No: 978-1-387-61173-7

THIS BOOK IS DEDICATED TO

SAMANTHA, MY HERO

WITH GRATITUDE TO:

My husband, for his unconditional love

My family, for their love and support

My friends, for always being there for me

Oprah, for her inspiration

Playboy magazine and Hugh Hefner, for the fun times

Jessica Bryan, my editor, and

Ian Kirkpatrick, my cover illustrator.

CONTENTS

Preface 1

Spiritual Affirmations 15

Chapter One: Gratitude is the Best Attitude 18

Chapter Two: Avoid Taking the Actions of
Others Personally 22

Chapter Three: The Opinion of the Mouse Does
Not Affect the Lion 27

Chapter Four: Your Problems Usually Begin
with Yourself 32

Chapter Five: The Universe Responds to Passion 37

Chapter Six: The Universe Gives Us More When
We Are Grateful 42

Chapter Seven: Who (or What) is the Universe? 46

Chapter Eight: Take a Deep Breath and Meditate 51

Chapter Nine: Happiness is a Choice, Not a Goal 55

Chapter Ten: It's a Perfect Event 59

Chapter Eleven: Get on a Good Frequency 62

Chapter Twelve: Remaining Whole When Others Want a Piece of You 67

Chapter Thirteen: Count Your Blessings 72

Chapter Fourteen: Tears of Joy 76

About the Author 84

PREFACE

The idea of living a more positive life began for me when I was about eighteen years old. My mind opened and shifted, and I began to work towards manifesting my dreams and goals.

At the time, all I really cared about was being in *Playboy*. I visualized it and even cut my picture out and pasted it on the cover of an existing issue of the magazine. I was trying to do anything that might help manifest my dream.

As the universe would have it, there was a casting call in Orlando, Florida. My best friend Brittany and I packed up and went to Orlando. I was nineteen years old, spray-tanned, bleached blond, and eager to conquer the world.

Waiting in the lobby of the hotel where the casting call was scheduled to take place, for some reason I wasn't at all nervous. When it was my turn, I stepped into the casting room and met the photographer, who is well known for his beautiful work. He put me in a simple pose wearing lingerie and took several amazing photos. After we left, I was not really sure what might come of it. I was just proud of myself for going and happy to have the company of my best friend. We were young and excited for the future.

We stopped to get a burger on our way home, and as we were eating, I received a telephone call asking me to be in Miami the next day. Evidentially my photos had been well received! It was the moment I had been visualizing, the moment I had waited for my whole life, the moment my dream started to become a reality.

By chance, my aunt Linda and grandma were in town, and when I returned home they expressed their support of my goal. They drove me down to the shoot and dropped me off personally. At the hotel where the shoot was going to be held, I felt great admiration for the staff and my good fortune. I tried to soak in each wonderful moment as much as I could.

The photo shoot was to be held the next day. Now...here is the best part. I had never done a modeling job prior to this – never had even one modeling photo taken. So that's right, ladies and gentlemen, my first photo shoot ever was for *Playboy*, and it led to my photos being published in two magazines. I was an official *Playboy* model.

Of course, I was so grateful, but I wanted the "whole enchilada." I wanted the whole experience. I wanted to be lounging by the pool at the Playboy Mansion, dining with Hugh Hefner ("Hef"), and enjoying the amazing lifestyle that came with being a Playmate.

I spent the next three years visualizing and using various techniques and tools to get to the next step. When

I was about twenty-one, I discovered a very important spiritual practice: gratitude. This changed everything. I literally fell in love with the practice of being grateful because I knew it was the best way to pursue my dreams.

Out of nowhere, I felt an impulse from the universe to write a letter to Hugh Hefner. My letter was simple and to the point: "Hey, I am a *Playboy* model and my dream is to visit the Playboy Mansion and meet you.

I sent the letter and went about my life. Then, three weeks later, I received an email from a woman in Hef's security department. She told me Hef loved my letter and he wanted to invite me to dinner at the Mansion.

This invitation was beyond my wildest imagination. It fulfilled my utmost desire and I was filled with gratitude. Next, the universe responded with everything I needed: enough airline points for the flight, and my family offered enough points to put me in a hotel. I was on my way to Los Angeles for the first time.

At the airport, I caught a cab, and soon we were rolling up the driveway to the Playboy Mansion. It was just as I had envisioned it. I opened the front door – just as I had imagined I would – and I was filled with amazement at how I had manifested this situation.

After getting settled in my room, I went downstairs to wait for the festivities to begin. Just then, Hef turned a corner and entered the room. I heard him say, "Where's my girl?" What a surprise it was to realize he was referring to me! We hugged and someone took a picture of us together.

Dinner that night was a delightful buffet with so many choices it made me delirious – there were five kinds of salad and dressings, tiny bits of fancy cheese and baby onions on toothpicks, bowls of caviar, aromatic bread, whole baked chickens, a variety of vegetables, and salmon filets swimming in butter and lemon. Of course, the menu also included a dozen different choices for dessert. There were three separate tables in the middle of the room and a small circle filled with people who appeared to be regular houseguests at the Mansion. Hef sat at the head of the table in the middle of the room, laughing and joking with his friends.

After dinner, everyone was invited to watch a movie in his home theater. It was a black and white movie filled with his close friends and family. Hef was a kind person who loved to take care of his guests and entertain them. Yes, he was a very kind man, and there is nothing to say about him and his company other than they were gracious and welcoming.

Life continued on in the same manner after I returned to Florida. After two more years had gone by, I was struck with another wonderful idea. I decided to be a painted lady! This meant being spray-painted and walking around during parties – it was a really good way to enjoy some of the most exciting parties you could ever imagine.

We were instructed to do a video submission, which included telling Hef that we were excited to be part of the painted lady project. Of course, I used my inner acting skills and made my own skit.

My mother had an old, worn out 1920s-type suitcase buried in the back of her closet – I have no clue why she kept it there, but she did – and I pulled it out for my video skit. I wound my hair up in tight curls and applied red lipstick. It was an attempt to play the part of a pin-up type girl who was eager to visit the Playboy Mansion and meet Hef. I wanted him to see me with my bag packed and ready to travel. In the video,

I posed with the suitcase in my hand and said something like, "I'm ready to go!"

Someone told me later that Hef loved the video and watched it several times. Guess what? I got the job! I was a painted lady on two separate occasions at the Mansion in Los Angeles. The first time, I was taken to the gym and my naked body was painted. My painted "outfit" looked just like real clothing. The party event was called "Midsummer Night's Dream." We danced all night and had a wonderful time.

The second time I went to Los Angeles, I was painted to look like a nymph. It was a Halloween party filled with celebrities and others people who filled me with awe. Incidentally, the paint was quite hard to scrub off.

The parties at the Playboy Mansion were always so well organized. Hef seemed to take great joy in making people happy. He was generous and thoughtful, and he really seemed to care about people having fun.

The story went on for another year, and I continued to be persistent about my desire to be a centerfold in *Playboy* magazine. I was not going to take "no" for an answer. I thought, *There is no way they can say "no" forever, right?*

Eventually, I was given an opportunity to do a test photo shoot to be a Playmate. Not only did I get to do the test shoot, I was invited to stay at the Playboy Mansion because I lived out of town. So…there I was, staying in the guesthouse at the Mansion living my dream, hanging out with the other young women, and attending the events going on at the time. Everything I had created from my inner mindset, from my thoughts, had become my reality.

The universe seemed to always direct me to the right people, always at exactly the right time. The photographer Danny Stein represented the true beginning my modeling career. He gave me the guidance and direction I had been searching for all along. I am so grateful for the support Danny and his wife have given me over the years. They have done so much for me, personally and professionally.

Danny took the photos for the test shoot. Of course, the universe had arranged our meeting, but why? Obviously, it was because he knew best how to help me manifest my dreams. Danny was a photographer for *Playboy*. Everyone loved the pictures he took of me, and before I even realized what was happening, I was a *Playboy* centerfold! After years of trying, I had reached my goal.

* * *

Following my experiences with Hugh Hefner and *Playboy*, I began seriously pursuing the study of spirituality. In this book, I have shared what I learned, including the practice of gratitude and using spiritual affirmations.

You will find suggested spiritual affirmations at the end of each chapter. Affirmations can help you manifest your dreams, and I encourage you to take up this practice in whatever way best serves you.

With Gratitude!

Kimberly Kingsbury

2018

SPIRITUAL AFFIRMATIONS

> According to the American English Dictionary, to "affirm" is to state that something is true.... Affirmations are dynamic and practical.
> – www.ananda.org

When you pray or meditate, include a simple affirmation for the day. Say it to yourself several times so it can work inside of you to produce beneficial results. This practice can strengthen your faith in God, no matter what formal religion you were raised in or what church you might currently attend. Affirmations can also help you stay focused and take positive action in your daily life.

Relax in a quiet place, contemplate your current concerns, and then simply repeat the affirmation of your choice. Do this once a day at the same time. You can use the examples at the end of each chapter or make up your own. Say the affirmation multiple times with sincerity and feeling so it can go deep within. You can speak quietly in your mind or speak out loud, but always make sure you state your affirmations in the present tense – as though your intention has already come true.

This practice has the power to transform your inner and outer life. The words and the energy behind them will align you with creative, healing, spiritual forces and help you achieve your goals. By clearly stating what you want to achieve, you can actually create your own reality and make your dreams come true.

Chapter One

Gratitude is the Best Attitude

Let's contemplate this for a moment: gratitude is the best attitude. It just feels right, doesn't it? Let's learn how to change our lives and allow more good and happy experiences to manifest. Let's figure out exactly what we want our lives to be. Everything is possible with gratitude. This is the one practice I have always used. Gratitude is my daily habit, and it has drastically changed my life. I do not have time to feel sorry for myself because I'm busy being grateful.

Many of us spend much of our time feeling sorry for ourselves. We really must learn to stop being so self-centered. Remember, every person in the world has something going on at all times that is important to them. We have the choice to react – or not react – to the actions of others.

For example, I have a huge family – I'm one of five siblings. So let's just say there is never a boring moment. There are times when I allow everything going on around me to affect my inner peace, but instead I remind myself to be grateful. I tell myself: *I am so grateful that his issue has resolved itself. I am so grateful she feels better and her health has been restored. I am so grateful that I can maintain peacefulness inside of myself.*

You see...life is going to happen. The universe will throw lessons at you each and every day, but if you respond with gratitude you can produce positive change. If you do not learn from a particular lesson the first time, you will have the same experience over and over again until you do learn.

So begin using gratefulness to assist you in making small changes that will shift your energy in a more positive direction.

AFFIRMATIONS FOR CHAPTER ONE

1. I will be grateful for everything that happens today.
2. I will remain at peace, no matter what happens.

Chapter Two

Avoid Taking the Actions of Others Personally

Sometimes we think people are going out their way to hurt us and we take their actions personally. The reality is that no one cares enough to intentionally harm us. People are too busy worrying about themselves and how to better their current circumstances.

So if your friend "Sally" forgot to call you on your birthday, it's not because she doesn't like you. It's because she is busy with her own interests. If someone has not talked to you in a while, don't think there is more to it than the simple fact that people are busy with their own concerns.

I have the greatest friends, some of whom I have known for as long as sixteen years – and I'm only twenty-eight so that's a huge part of my life. There are times when I don't see or even talk to one of my friends for months. Do I worry that I need to water my friendships to maintain them? No. I understand that no matter what happens we are still going to be friends. I know we are all adults and we are busy with working, going to school, and raising families. Some are buying houses and – as I write this – I am in the middle of planning my wedding.

Everyone is so busy! Does this mean we should not follow-up with our friends? No! I'm saying we shouldn't take it personal if people do not always behave as we want or expect them to. For example, if someone cancels a lunch date, tell yourself: *Oh, that's great, because now I have time to do something else. We can always catch up another time.*

How you react to people and situations is what really matters. Whatever happens, respond in an appropriate and positive way, rather than reacting negatively. Tell yourself: *This must be for my benefit. Whatever I expected is not supposed to happen today.*

Your life will be so much easier when you have this kind of attitude.

AFFIRMATIONS FOR CHAPTER TWO

1. The people in my life love and support me.

2. I take nothing personal no matter what others say or do.

Chapter Three

The Opinion of the Mouse Does Not Affect the Lion

The title of this chapter is one of my most favorite sayings. Sometimes it seems like everyone is guilty of worrying about the opinions of other people. This is human nature. Whether you believe it or not, you are constantly seeking approval from everyone around you. For example, you care about whether you might have said something upsetting to your mother, and you are scared you might have hurt her feelings. Your mother, on the other hand, might be having wonderful thoughts and feelings about you and your behavior.

You don't realize it, but you care. You buy clothes so people will compliment you. Perhaps you spent your time washing your car on Friday nights when you were in high school because you cared about what people thought of your car – and therefore what they might think about you. This type of mindset is so common. However, becoming aware of it can be the first step towards becoming awake and aware.

The point I am trying to make is that the opinion of the mouse does not affect the lion. When we tell ourselves that other people are better than us, smarter, richer, or more beautiful than us, we let these opinions affect our mood and happiness level. We need to be more objective and tell ourselves: *This is only their opinion, their viewpoint of the situation. It's not my truth.*

Personally, I don't make any effort to try and impress others – I am too busy trying to impress God to worry about what other people might think about me or anything else.

The best course of action is to *learn* from the opinions of other people. Contemplate what they have to say and discover what you can learn from them. It is possible to learn from almost any person or situation, no matter how difficult. Take a step back and look at all of the circumstances, and then consider every event as a classroom that offers you an opportunity to grow spiritually. If you live like this, your life will be so much easier and your relationships more rewarding.

AFFIRMATIONS FOR CHAPTER THREE

1. The opinions of others do not disturb me.

2. I accept myself exactly as I am.

3. My mind is at peace.

4. I evaluate every situation without judgment.

Chapter Four

Your Problems Usually Begin with Yourself

Repeat after me: *I am causing my own problems.* Now reflect on what you can do to change your mood and be in a better frame of mind so you are not arguing with everyone.

This is big. If you can do this, you can conquer the world. People love to blame others: their parents, their friends, their spouse, and their employer. I have wasted so much of my life listening to other people complain about what one person did to some other person. Now I just laugh because I know the truth.

There are times when I might be in a small fight with more than one person. I am talking little arguments. You know, those days when you feel like everyone around you is being difficult and you think the whole world is against you. Yes, my friends, I'm talking about those kinds of days.

My point in all of this is that we need answers. Are you sick of the broken relationships you have with certain people? Do you want to live in such a way that you are able to self-reflect and own up to your faults, learn from them, and become a happier and more fulfilled person?

Let's face it: who doesn't? Just let go of your ego. Right this very moment! Your ego is telling you that your opinion is right, and that this person did this, and that person did that. But none of this sort of blaming makes any sense.

If you are arguing with more than one person in your life, recognize your own actions, take responsibility, and change your behavior. Reflect on how you can better yourself as a person. This will be different for each of us, so it has to be custom-tailored to each individual. The fact of the matter is that we are all aware of our negative behaviors. We all know what we need to do to improve our relationships. So decide to make positive changes in the way you relate to others, and also to yourself.

AFFIRMATIONS FOR CHAPTER FOUR

1. I do not blame anyone including myself.

2. I take responsibility for my actions.

Chapter Five

The Universe Responds to Passion

What lights up your soul? What puts you into a total state of bliss? Most people spend their whole lives trying to figure this out. I am lucky enough to have found my personal passion.

My family has a home movie that shows me at age four. I'm standing in the backyard of the house I grew up in, and there are trees behind me lining the fence. It's a majestic, jungle-like place. There is also a man-made canal behind me filled with unknown, possibly dangerous possibilities.

As I look at this movie, I can see that my four-year-old self had already learned how to pose for the camera. The child I had been knew her passion. At the time, I recall saying to my grandmother, "Grandma! Videotape me. I want to make a movie. I want to be an actress."

During my teenage years, my desire turned towards the modeling world, and I became deeply fascinated with that industry. This eventually led to my photos appearing in *Playboy* magazine as a featured centerfold Playmate. It was a long journey and there were many hoops to go through. To finally get there six years later I had to meet so many people and learn so many lessons to reach the fulfillment of my dreams. I can tell you one thing for certain: if you are passionate enough, you will do anything necessary to achieve your goals.

Looking back, I am grateful that I had this passion because it led me to the final realization that my truest passion is God, my higher power. Now I spend most of my time feeding my consciousness with more and more knowledge about the nature of the universe. I love learning and I am madly in love with the quest for spiritual enlightenment!

Passion should be effortless. The universe responds to passion. If you are passionate about something it will be like a musical note that rings out from you and spreads like the wind, effortless, charming, and just right. The universe will listen and grant your wildest, most precious desires.

AFFIRMATIONS FOR CHAPTER FIVE

1. The universe gives me love and support.

2. God and all of creation light up my soul.

3. My deepest dreams are already fulfilled.

Chapter Six

The Universe Gives Us More When We Are Grateful

This is a simple law. Let's say I took you out for lunch and it felt good enough to be your birthday lunch. I am talking oysters and a nice Caesar salad, followed by a fish filet. Would expect me to take you back to a restaurant again if you did not say "thank you?"

I highly doubt it. If you do not show gratitude to others, especially the universe, you are in big trouble. You see...what happens to most people is they do not even realize how ungrateful they are. They get the five dollars they need so badly to buy a train ticket, but they don't express gratitude to the person (and the universe, in general) who gave them the money.

This is a problem. The universe will never give you more of anything if you do not say "thank you." Expressing gratitude allows the universe to give you more of what you need. For example, you really like your job and the work you do. If you want to continue having more enjoyable experiences and receiving a good salary, simply express your gratitude. Another example: Because I might want my agent to book more photo shoots, I always remember to thank him for his efforts on my behalf.

I hate to go into the negative, but I think people need to be more aware of this basic principle. If you are not grateful, the universe will take your happiness and prosperity away from you, slowly but surely. I don't care if you are so rich that you cannot spend your money fast enough. The universe will take it all from you and give it to someone who is grateful. This is universal law. So let's start with being grateful for everything we receive, knowing there always someone with less and always a reason to value what we have been given.

AFFIRMATIONS FOR CHAPTER SIX

1. I am grateful for everything I receive.

2. The abundance of the universe flows easily into my life.

Chapter Seven

Who (or What) is the Universe?

I grew up Christian so I believe in this religion. I spent my whole life practicing the beliefs of my faith. I devoted much of my time to being active in my church when I lived in Los Angeles, and I was even re-baptized. My husband is Jewish and I am open to learning about his faith, while also keeping my beliefs in my heart.

When I first started college, I took a class on the religions of the world. One thing I learned about all the different religions is that every single one of them holds a common belief in the idea there is a higher being, a higher "power." So my point is that the "universe" can be anything that you feel is your higher power. Simply stated, my universe is God. So when I refer to the universe, I am referring to God. Of course, this might be different for different individuals.

My point is that over the years there has been huge confusion as to the nature of the spiritual universe. Some people think the New Age folks have made up their own higher power and it is revolutionizing the world.

Yes, this is true, but it's because so many of us have awakened our consciousness, our awareness. We are starting to see the truth, but this doesn't mean we have made up our own religion. It's just a basis that anyone can relate to.

In my study of religions, I learned something very interesting. Jesus himself has often been portrayed with a white light, or halo, radiating around his head. This illustrates that Jesus was connected to God thru this white, golden light.

There is a practice you can do in Reiki in which you draw the white golden light down into your Seventh Chakra, which is the energy center above the top of your head. So my point is that everything is connected, and the universe is trying to show you that religion is connected to New Age concepts and beliefs.

I find it reassuring to know that the universe will support me when I am in a positive frequency. The universe will show you different indicators in your everyday life to reassure you that you are going in the right direction for your spiritual growth.

For example, have you considered a specific subject or a vacation destination and suddenly it seems that everywhere you look the universe is giving you signs and situations that mirror what you are thinking about? This is the universe helping you reach your goals and reassuring you that are heading in the right direction.

AFFIRMATIONS FOR CHAPTER SEVEN

1. I trust the universe to lead me in my spiritual growth.

2. I am connected to my higher power.

3. I am safe and reassured by the universe.

Chapter Eight

Take a Deep Breath and Meditate

Take a deep breath in and a deep breath out. Breathing keeps the body alive and helps us stay connected to what is going on outside and inside of us. Breathing – the breath of life – is connected to the central nervous system and the oxygen it provides feeds the entire body. Following the breath meditation reminds us to be still, to be present. Meditating means taking time for your personal development. It's a process that will help you learn about yourself and the world around you. Meditation takes us out of our everyday shuffle. It's an inner place where we can be still and connect to God. When you meditate, listen to his guidance. Tell

God what you want to create, and then simply breathe with conscious awareness. With this practice, you will receive everything.

My friend, Deb, who is also my spiritual teacher, taught me another avenue of discovery. Try doing this form of meditation when you are walking around and not actively engaged in anything else. Repeat these words over and over quietly to yourself: "So-Ham." This is a Hindu mantra meaning "I am That." Alternatively, you can simply say, "I am." Repeating this mantra will connect you with the universe, your higher power.

Now think of something you would like to create in your life. Tell the universe your hopes and dreams in a way that allows the universe to know you are ready to receive. Your spiritual and worldly life will benefit greatly from the establishment of a regular meditation and affirmation practice.

AFFIRMATIONS FOR CHAPTER EIGHT

1. I meet the universe inside of myself during meditation.

2. All of my dreams are manifest right now!

3. I am one with God.

4. I meet my true self between the in-breath and the out-breath.

Chapter Nine

Happiness is a Choice, Not a Goal

Happiness is a choice. Whether we realize it or not, we make the choice to be happy or sad, depressed or angry all the time. You can choose to react to any situation or person in a negative or positive way. Most of us want to be happy, but sometimes we are not because of how we react to those around us.

I am faced with this problem daily, and I'm sure other people are, too. Because we have free well, we can decide whether we are going to *react* and let other people and situations ruin our happiness.

We also have the option to simply *respond* without getting into negativity.

For example, perhaps you are mad at your wife because she didn't answer the phone when you called. This is a choice. Maybe you are jealous of someone who has a beautiful new outfit. This is also a choice. If you want to make a fool of yourself, that's a choice.

Do you see my point? I am not saying that life won't throw curve balls at you all day long. It's just that we need to maintain our own state of happiness. This means taking responsibility and not letting the daily drama of life affect us.

Many people have told me I live in a fantasy world, and that my life is all rainbows and butterflies. I guess I come off as a true optimist. The truth is I do not want to be any different. I love being in a good mood. I love letting everything be all positives rather than negatives. I love choosing to be happy, and the more I make this choice, the happier I become!

So... do you want to be an optimist or a pessimist? It's your choice.

AFFIRMATIONS FOR CHAPTER NINE

1. Today I choose to be happy.

2. I am peace with the people and events around me.

Chapter Ten

It's a Perfect Event

Something wise and wonderful happened that changed my life. As I wrote about in the Preface to this book, a series of wonderful events led me to living in Los Angeles. At that point, I was ten years into my expanded way of thinking and was in the presence of great thinkers. During each chapter of my life, I have attracted the perfect people to teach me the lessons I need to learn.

The most important lesson we all need to learn is that everything – whether good or bad – happens for our own benefit and spiritual growth. Each event is a *per-*

fect event. This has been proved to me over and over again. It doesn't matter the situation, something good always comes from adversity.

This might mean growing from losing a baseball game, finding true love after a divorce, or working at a job you are passionate about – after you have been fired from a job you didn't care about very much. This, my friend, is a perfect event. Remember, everything that happens to you is for your benefit and spiritual growth.

AFFIRMATIONS FOR CHAPTER TEN

1. Everything happens for a reason.

2. My life is perfectly ordered by a higher power.

3. The universe is leading me to perfection.

Chapter Eleven

Get on a Good Frequency

You have probably heard people say: *Get on my level!* This is my way of thinking – let's all get together on the best frequency. Every day, I try to keep myself feeling good and positive. I refuse to let outside circumstances effect my inner lion.

So...how do I know when I am on a good frequency? Let's find out by creating our own good "frequency reminder." First, pick an object, any object. Make it some you love, an animal, a food, a tree, or a flower. Once you have decided on your frequency reminder, whenever you see it you will automatically become aware that you are on a good frequency with the universe. Even now, butterflies often land right in front me.

A while ago, I was living with my two best friends, Samantha and Cheriee. We would get together and talk about our lives and feed our spirituality daily. I taught them about getting on a good frequency and picking a sign that would help. We each picked a sign, and every time we saw our sign we would know we were on a good frequency with the universe.

My sign was a butterfly; Samantha picked a sunflower; and Cheriee picked a hummingbird. I explained that when we were separated. If we saw one of the other's sign, it meant we were together on an uplifting frequency. Even now, butterflies often land near me.

Time passed, and after I moved to Malibu hummingbirds would often flutter right in front of my window. I knew it was the universe reassuring me that I was on a good frequency with the universe – and with Cheriee, who had chosen a hummingbird as her sign.

Sadly, Samantha passed away. The day after her death, I had to fly to Chicago. Although I was grief-stricken, in the midst of my grief I found happiness. As I was driving home from the airport, I looked out of the window and saw a field of glorious, wild, yellow sunflowers dancing in the wind. Now...I am from Florida, and you don't just see wild sunflowers out your window in Florida! It was Samantha's way of telling me from heaven that everything was okay and that she was with me in spirit.

You will actually start to notice the sign you chose everywhere, and you will be filled with bliss every time you see it. You will experience the beginning of the magical connection we all have with the universe. So...what's your universal sign going to be?

AFFIRMATIONS FOR CHAPTER ELEVEN

1. Today I will remember my frequency reminder and be joyful.

2. Those I love are always with me, even when they are not physically present.

Chapter Twelve

Remaining Whole When Others Want a Piece of You

In speaking with my brother, we got on the topic of how it can be difficult to live a peaceful spiritual life when we have to deal with the stresses of everyday life. Spiritual people still have to go to work every day to earn a living; some have kids or relationships that can be problematic. This is a vital question for anyone who is on the spiritual path. We are all learning how to remain at peace in the midst of life, which can be chaotic.

Much of my day is devoted to seeking enlightenment through meditating and studying different spiritual practices. I might have my salt lamps turned on, my aromatherapy going, and meditative music playing, and then BAM! a family member who is not on the same frequency rings me on the phone. This used to be extremely stressful because I did not want to hear from anyone who was not absolutely positive. I experienced a couple of years during which I kept the phone turned off. After all, if I did not have a phone, no one would be able to shift my nice, positive attitude.

Keeping the phone turned off worked, but it was not realistic. I mean... I lived months at a time without even having a cell phone. At one point, I didn't have a phone for six months, but not because I couldn't have one. I just wanted to be able to maintain a positive mental state all day without talking to other people, who might put me in a troubled mood.

Maintaining this attitude is still a work in progress. What I have found to be most effective is coming to the realization that life is going to happen. Your baby will throw up on you, or you might be late for work. You might have a friend or family member who calls to unload their drama. The point is that these things don't really matter.

The key is that everything is perfect. Events are supposed to happen exactly the way they do. For example, instead of getting stressed out, tell yourself: *I am grateful that I lost my job today, because now a better opportunity is coming my way.*

Listening to other people's drama, I realize my world is not like theirs. Sometimes when I get off the phone, I am so grateful that I do not let the energy of other people affect me. I affirm that my life situation is not similar to theirs. I simply center myself in the frequency of the universe and send them my love and blessings. This is how you can give other people a piece of yourself, but still remain whole.

AFFIRMATIONS FOR CHAPTER TWELVE

1. I am independent of the opinions and actions of others.

2. I accept whatever happens with peace in my heart.

3. I radiate love to all.

Chapter Thirteen

Count Your Blessings

Many people like to acquire things. Do not allow these desires to consume you. If you have a strong desire for material possessions, take a step back and ground yourself. Maybe you can go to the beach and put your feet in the sand – really get reconnected to the earth.

On the other hand, you don't need to feel guilty about your possessions, nor do you need to be concerned that you have more wealth than others. Actually, if you are ungrateful for what the universe has given you, guess what? The universe will take it right back.

You have probably heard this before: count your blessings. Be grateful for everything you have. I can tell you personably how gratitude has changed my life. Several years ago, when I began to understand the power of gratitude, I started writing letters. At first, I wrote letters to family members thanking them and letting them know how much they are appreciated.

Now...when I come into contact with the public – even if it is only the barista who did a good job serving my coffee – I go out of my way to write a letter to the corporate office and tell their supervisor about what a good job the employee did.

AFFIRMATIONS FOR CHAPTER THIRTEEN

1. I am grateful for my abundance but I am not attached to it.

2. I am grateful for every experience.

Chapter Fourteen

Tears of Joy

My Sunday nights often consist of relaxation, enjoying my husband's company, experiencing unconditional love with my dogs, and, of course, Oprah's *Super Soul Sundays*. I always watch this show because it helps me connect to the deeper meaning in my experiences and the world around me. On her show, Oprah interviews many New Age authors and spiritual leaders, as well as experts in the area of health. I just love it.

One particular Sunday was especially meaningful. I found myself taking notes and pausing to repeat the moments when something struck me as important to make note of. The one-hour program turned into two hours of beautiful, spiritually-focused, meaningful conversation between my husband and I. We took a look at our everyday actions, thoughts, words, and career moves. We tried to look at our life together in a different way – the way God wanted us to look at it.

This was a life-changing moment for us. After the program was over, my husband looked at me and started crying. Mind you…I had never seen him cry before. He thanked me for opening up his heart to love, and it was a wonderful moment that really made me feel at ease in my soul. It felt like we were spiritually intertwined with each other, and with God.

Just then, he received an email that offered constructive criticism about a particular company we were interested in. It referenced the same article we had just read that morning, and we were already planning to follow through with this company. All the points made in the email were exactly what my husband had expressed to me earlier. The universe had tried to tell him the same thing several times, but it took this exact moment for everything to come to light for him and make sense.

We believe this particular email message was a blessing, a significant reminder that my husband was operating on the right frequency, because it had arrived when his energy was so high.

When I went for my nightly bath after the show, I took my notes with me and began the spiritual practice I had just learned from the guest Oprah had interviewed. Almost immediately, I began to cry tears of joy. At that moment, I felt like I had reached the next step in my spiritual growth. I came to the realization of how beautiful it is that we are able to learn these expanding spiritual concepts and practice them in our lives. How lucky we are to be so close to God. It's simply magical and magnificent.

We are never done learning. We just have to keep tuning into our higher power and embracing the spiritual lessons God puts before us.

AFFIRMATIONS FOR CHAPTER FOURTEEN

1. I embrace the lessons God gives me.

2. I will do as my higher power asks.

3. My eyes are open to the magic that is all around me.

Getting ready at the *Playboy* Mansion, for the mid-summers night dream party.

This is the night where I first met Hugh Hefner at this house where I had my first dinner and movie at his house.

This photo was taken in the backyard at the *Playboy* Mansion, right after I did my test shoot.

ABOUT THE AUTHOR

Kim Kingsbury believes in sharing her passion for life and spiritual growth with everyone she meets. She really enjoys sharing her knowledge of how everyday events can be positive and fulfilling just by looking at them differently. Kim's other passion is related to her experiences with *Playboy* magazine and Hugh Hefner at the Playboy Mansion. In this book, she has combined these two passions to illustrate how anyone can manifest his or her dreams.

Kim did commercial modeling and worldwide swimsuit competitions for ten years. More recently, she is focused on acting in movies and commercials, and following whatever path the universe has for her.

Kim is married to an amazing man who fills her life with joy and shares her deep understanding of how the universe works.